What Comes Before / After?

Kindergarten / First Grade

Thinking Skill Builder

Jenny Pearson

What Comes Before / After?

Kindergarten / First Grade Thinking Skill Builder

Jenny Pearson

Kivett Publishing

ISBN: 978-1-941691-43-4

Juvenile > Concepts > Counting & Numbers

Juvenile > Concepts > Alphabet

TABLE OF CONTENTS

INTRODUCTION

Memorizing a list—like the alphabet or numbers—gives you valuable knowledge. Being able to determine which letter or number comes next is a valuable thinking skill. Suppose that you are trying to visit a friend who lives in a particular apartment, and you just passed apartment K. If the apartments are in order from A to Z, knowing which apartment comes next may be helpful. If you need to start saying the alphabet from A, although you have memorized the alphabet, you're lacking a valuable thinking skill. Being able to quickly determine which letter comes after K (for example) is a valuable thinking skill.

This workbook was prepared with two primary goals:

- Offer practice with kindergarten / 1st grade facts.
- Help young students develop their thinking skills.

1

UPPERCASE LETTERS
A to Z

A B C D E F G H I J K L M
N O P Q R S T U V W X Y Z

1. Which letter comes after uppercase E?

2. Which letter comes after uppercase R?

3. Which letter comes after uppercase J?

4. Which letter comes after uppercase X?

5. Which letter comes after uppercase C?

6. Which letter comes after uppercase N?

7. Which letter comes after uppercase I?

8. Which letter comes after uppercase U?

9. Which letter comes after uppercase P?

10. Which letter comes after uppercase D?

11. Which letter comes after uppercase K?

12. Which letter comes after uppercase Y?

13. Which letter comes after uppercase S?

14. Which letter comes after uppercase M?

15. Which letter comes after uppercase V?

16. Which letter comes after uppercase G?

17. Which letter comes after uppercase O?

18. Which letter comes after uppercase F?

19. Which letter comes after uppercase W?

20. Which letter comes after uppercase H?

21. Which letter comes after uppercase T?

22. Which letter comes after uppercase Q?

23. Which letter comes after uppercase B?

24. Which letter comes after uppercase L?

25. Which letter comes before uppercase X?

26. Which letter comes before uppercase G?

27. Which letter comes before uppercase P?

28. Which letter comes before uppercase J?

29. Which letter comes before uppercase V?

30. Which letter comes before uppercase S?

31. Which letter comes before uppercase N?

32. Which letter comes before uppercase Z?

33. Which letter comes before uppercase R?

34. Which letter comes before uppercase C?

35. Which letter comes before uppercase M?

36. Which letter comes before uppercase H?

37. Which letter comes before uppercase L?

38. Which letter comes before uppercase T?

39. Which letter comes before uppercase I?

40. Which letter comes before uppercase Y?

2

NUMBERS
1-100

1 2 3 4 5 6 7 8 9 10

10 20 30 40 50 60 70 80 90 100

1. Which number comes after 6?

2. Which number comes after 42?

3. Which number comes after 91?

4. Which number comes after 17?

5. Which number comes after 29?

6. Which number comes after 78?

7. Which number comes after 55?

8. Which number comes after 60?

9. Which number comes after 14?

10. Which number comes after 39?

11. Which number comes after 87?

12. Which number comes after 68?

13. Which number comes after 40?

14. Which number comes after 53?

15. Which number comes after 25?

16. Which number comes after 99?

17. Which number comes after 63?

18. Which number comes after 44?

19. Which number comes after 56?

20. Which number comes after 19?

21. Which number comes after 28?

22. Which number comes after 71?

23. Which number comes after 35?

24. Which number comes after 92?

25. Which number comes before 8?

26. Which number comes before 32?

27. Which number comes before 77?

28. Which number comes before 26?

29. Which number comes before 49?

30. Which number comes before 60?

31. Which number comes before 81?

32. Which number comes before 13?

33. Which number comes before 61?

34. Which number comes before 30?

35. Which number comes before 89?

36. Which number comes before 48?

37. Which number comes before 27?

38. Which number comes before 100?

39. Which number comes before 6?

40. Which number comes before 75?

3

LOWERCASE LETTERS

a to z

a b c d e f g h i j k l m

n o p q r s t u v w x y z

1. Which letter comes after lowercase k?

2. Which letter comes after lowercase d?

3. Which letter comes after lowercase y?

4. Which letter comes after lowercase j?

5. Which letter comes after lowercase o?

6. Which letter comes after lowercase f?

7. Which letter comes after lowercase a?

8. Which letter comes after lowercase q?

9. Which letter comes after lowercase l?

10. Which letter comes after lowercase g?

11. Which letter comes after lowercase p?

12. Which letter comes after lowercase x?

13. Which letter comes after lowercase b?

14. Which letter comes after lowercase s?

15. Which letter comes after lowercase i?

16. Which letter comes after lowercase t?

17. Which letter comes after lowercase c?

18. Which letter comes after lowercase r?

19. Which letter comes after lowercase h?

20. Which letter comes after lowercase n?

21. Which letter comes after lowercase u?

22. Which letter comes after lowercase w?

23. Which letter comes after lowercase m?

24. Which letter comes after lowercase e?

25. Which letter comes before lowercase v?

26. Which letter comes before lowercase k?

27. Which letter comes before lowercase s?

28. Which letter comes before lowercase i?

29. Which letter comes before lowercase g?

30. Which letter comes before lowercase y?

31. Which letter comes before lowercase d?

32. Which letter comes before lowercase o?

33. Which letter comes before lowercase f?

34. Which letter comes before lowercase t?

35. Which letter comes before lowercase p?

36. Which letter comes before lowercase l?

37. Which letter comes before lowercase q?

38. Which letter comes before lowercase j?

39. Which letter comes before lowercase z?

40. Which letter comes before lowercase x?

4

BIG NUMBER CHALLENGE (100+)

100 200 300 400 500 600

700 800 900 1000 1100 1200

1. Which number comes after 123?

2. Which number comes after 654?

3. Which number comes after 1,207?

4. Which number comes after 256?

5. Which number comes after 199?

6. Which number comes after 348?

7. Which number comes after 300?

8. Which number comes after 109?

9. Which number comes after 945?

10. Which number comes after 129?

11. Which number comes after 475?

12. Which number comes after 700?

13. Which number comes after 333?

14. Which number comes after 811?

15. Which number comes after 599?

16. Which number comes after 2,057?

17. Which number comes after 144?

18. Which number comes after 719?

19. Which number comes after 3,000?

20. Which number comes after 328?

21. Which number comes after 530?

22. Which number comes after 9,999?

23. Which number comes after 1,999?

24. Which number comes after 1,099?

25. Which number comes before 243?

26. Which number comes before 899?

27. Which number comes before 528?

28. Which number comes before 300?

29. Which number comes before 1,234?

30. Which number comes before 777?

31. Which number comes before 4,000?

32. Which number comes before 326?

33. Which number comes before 110?

34. Which number comes before 478?

35. Which number comes before 576?

36. Which number comes before 270?

37. Which number comes before 219?

38. Which number comes before 100?

39. Which number comes before 987?

40. Which number comes before 1,500?

5

NUMBER
WORDS

one two three four five six

seven eight nine ten eleven

1. Which number comes after seven?

2. Which number comes after fifty-six?

3. Which number comes after twenty?

4. Which number comes after ninety-five?

5. Which number comes after eighteen?

6. Which number comes after eleven?

7. Which number comes after forty-three?

8. Which number comes after thirty-nine?

9. Which number comes after sixteen?

10. Which number comes after thirty-five?

11. Which number comes after ninety-two?

12. Which number comes after seventy-one?

13. Which number comes after forty-seven?

14. Which number comes after nineteen?

15. Which number comes after fifty-eight?

16. Which number comes after twenty-four?

17. Which number comes after thirty-four?

18. Which number comes after six?

19. Which number comes after twenty-two?

20. Which number comes after seventy-six?

21. Which number comes after fifty?

22. Which number comes after ninety-eight?

23. Which number comes after thirteen?

24. Which number comes after seventy-nine?

25. Which number comes before nine?

26. Which number comes before ninety?

27. Which number comes before sixty-three?

28. Which number comes before eleven?

29. Which number comes before twenty-eight?

30. Which number comes before ten?

31. Which number comes before eighty-seven?

32. Which number comes before thirty-six?

33. Which number comes before forty-two?

34. Which number comes before twelve?

35. Which number comes before seventy-three?

36. Which number comes before thirty?

37. Which number comes before seven?

38. Which number comes before fifty-one?

39. Which number comes before one hundred?

40. Which number comes before one?

6

DAYS OF
THE WEEK

Sunday Monday Tuesday

Wednesday Thursday

Friday Saturday

Practice writing the days of the week in order.

Note that <u>Wednesday</u> has an 'n' after the first 'd.'

Sunday Monday Tuesday Wednesday

Thursday Friday Saturday

_____

_____

_____

_____

_____

_____

_____

_____

_____

_____

_____

_____

_____

Try to memorize the days of the week.

1. Which day comes after Tuesday?

2. Which day comes after Friday?

3. Which day comes after Thursday?

4. Which day comes after Sunday?

5. Which day comes after Wednesday?

6. Which day comes after Monday?

7. Which day comes after Saturday?

8. Which day comes after Friday?

9. Which day comes after Sunday?

10. Which day comes after Thursday?

11. Which day comes after Saturday?

12. Which day comes after Wednesday?

13. Which day comes after Tuesday?

14. Which day comes after Monday?

15. Which day comes before Wednesday?

16. Which day comes before Monday?

17. Which day comes before Thursday?

18. Which day comes before Sunday?

19. Which day comes before Friday?

20. Which day comes before Tuesday?

21. Which day comes before Saturday?

22. Which day comes before Thursday?

23. Which day comes before Monday?

24. Which day comes before Saturday?

25. Which day comes before Tuesday?

26. Which day comes before Friday?

27. Which day comes before Sunday?

28. Which day comes before Wednesday?

7

MONTHS OF
THE YEAR

January February March

April May June

July August September

October November December

Practice writing the months of the year in order.

Note that <u>February</u> has an 'r' after the 'b.'

January February March April May June July

August September October November December

_____

_____

_____

_____

_____

_____

_____

_____

_____

_____

_____

_____

_____

_____

Try to memorize the months of the year.

1. Which month comes after March?

2. Which month comes after June?

3. Which month comes after September?

4. Which month comes after February?

5. Which month comes after July?

6. Which month comes after August?

7. Which month comes after November?

8. Which month comes after May?

9. Which month comes after January?

10. Which month comes after October?

11. Which month comes after April?

12. Which month comes after December?

13. Which month comes after February?

14. Which month comes after July?

15. Which month comes after September?

16. Which month comes after August?

17. Which month comes before June?

18. Which month comes before November?

19. Which month comes before February?

20. Which month comes before May?

21. Which month comes before January?

22. Which month comes before April?

23. Which month comes before December?

24. Which month comes before March?

25. Which month comes before July?

26. Which month comes before April?

27. Which month comes before September?

28. Which month comes before November?

29. Which month comes before August?

30. Which month comes before March?

31. Which month comes before October?

32. Which month comes before May?

8

COLORS OF
THE RAINBOW

red orange yellow green

blue indigo violet

Practice writing the colors of the rainbow in order.

Red appears at the top, with violet at the bottom.

The first letters of each color spell Roy G. Biv.

(Note: Not all textbooks include indigo.)

red orange yellow green
blue indigo violet

_____

_____

_____

_____

_____

_____

_____

_____

_____

_____

Try to memorize the colors of the rainbow.

1. Which color comes after orange?

2. Which color comes after green?

3. Which color comes after yellow?

4. Which color comes at the top of the rainbow?

5. Which color comes after indigo?

6. Which color comes after red?

7. Which color comes after blue?

8. Which color comes after yellow?

9. Which color comes after orange?

10. Which color comes after green?

11. Which color comes after red?

12. Which color comes after indigo?

13. Which color comes at the top of the rainbow?

14. Which color comes after blue?

15. Which color comes before green?

16. Which color comes before violet?

17. Which color comes before blue?

18. Which color comes before yellow?

19. Which color comes before indigo?

20. Which color comes before orange?

21. Which color comes at the bottom of the rainbow?

22. Which color comes before indigo?

23. Which color comes before blue?

24. Which color comes at the bottom of the rainbow?

25. Which color comes before yellow?

26. Which color comes before violet?

27. Which color comes before orange?

28. Which color comes before green?

9

PLANETS OF THE SOLAR SYSTEM

Mercury Venus Earth

Mars Jupiter Saturn

Uranus Neptune

Practice writing the planets of our solar system in order from the sun (beginning with Mercury).

Note that Pluto is currently considered to be a dwarf planet (and not one of the main planets).

Tip: Saturn Uranus Neptune (first letters = SUN).

Mercury Venus Earth Mars

Jupiter Saturn Uranus Neptune

_____

_____

_____

_____

_____

_____

_____

_____

_____

Try to memorize the planets of the solar system.

1. Which planet comes after Earth?

2. Which planet comes after Saturn?

3. Which planet comes after Mercury?

4. Which planet comes after Jupiter?

5. Which planet comes after Uranus?

6. Which planet comes after Venus?

7. Which planet comes after Mars?

8. Which planet is closest to the sun?

9. Which planet comes after Mars?

10. Which planet comes after Jupiter?

11. Which planet comes after Venus?

12. Which planet is closest to the sun?

13. Which planet comes after Earth?

14. Which planet comes after Uranus?

15. Which planet comes after Mercury?

16. Which planet comes after Saturn?

17. Which planet comes before Jupiter?

18. Which planet comes before Venus?

19. Which planet is furthest from the sun?

20. Which planet comes before Uranus?

21. Which planet comes before Mars?

22. Which planet comes before Saturn?

23. Which planet comes before Earth?

24. Which planet comes before Neptune?

25. Which planet comes before Saturn?

26. Which planet comes before Neptune?

27. Which planet comes before Earth?

28. Which planet comes before Jupiter?

29. Which planet comes before Venus?

30. Which planet is furthest from the sun?

31. Which planet comes before Mars?

32. Which planet comes before Uranus?

10

POLYGONS (SHAPES WITH STRAIGHT EDGES)

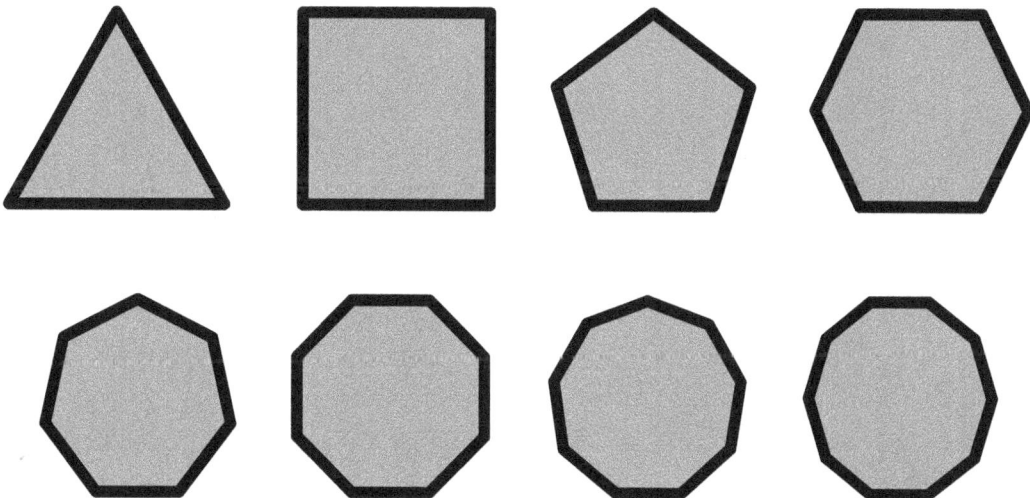

Trace the edges of the following polygons.

triangle
3 sides

square
4 sides

pentagon
5 sides

hexagon
6 sides

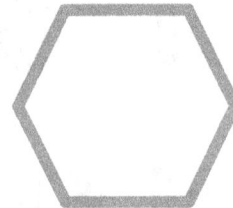

heptagon
7 sides

octagon
8 sides

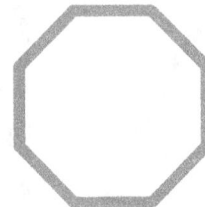

nonagon
9 sides

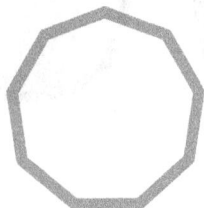

decagon
10 sides

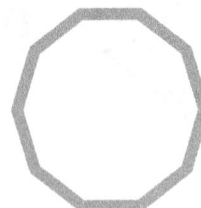

1. Name and draw the polygon that has one more side than a square.

2. Name and draw the polygon that has one more side than a hexagon.

3. Name and draw the polygon that has one more side than a triangle.

4. Name and draw the polygon that has one more side than a pentagon.

5. Name and draw the polygon that has one more side than a heptagon.

6. Name and draw the polygon that has one more side than a nonagon.

7. Name and draw the polygon that has one more side than a triangle.

8. Name and draw the polygon that has one more side than an octagon.

9. Name and draw the polygon that has one more side than a pentagon.

10. Name and draw the polygon that has one more side than a square.

11. Name and draw the polygon that has one less side than a hexagon.

12. Name and draw the polygon that has one less side than a square.

13. Name and draw the polygon that has one less side than a nonagon.

14. Name and draw the polygon that has one less side than a pentagon.

15. Name and draw the polygon that has one less side than a heptagon.

16. Name and draw the polygon that has one less side than an octagon.

17. Name and draw the polygon that has one less side than a decagon.

18. Name and draw the polygon that has one less side than a pentagon.

19. Name and draw the polygon that has one less side than a square.

20. Name and draw the polygon that has one less side than a hexagon.

11

SORTING NUMBERS

1 10 100 1,000 10,000

10 20 30 40 50 60 70 80

Example: Arrange the following numbers in order from smallest to greatest.

24	18	5	21	32	7	40	36

→

Solution: The smallest number is 5. Next comes 7, then 18, then 21, then 24, then 32, then 36, and 40. Enter these numbers in the bottom row. (We did this step for you in this example.)

24	18	5	21	32	7	40	36
5	7	18	21	24	32	36	40

→

Tip: Cross out each number on the top row after you write it in the bottom row. This will help you keep track of which numbers you have already used.

1. Arrange the following numbers in order from smallest to greatest.

8	13	7	2	9	16	4	15

→

2. Arrange the following numbers in order from smallest to greatest.

43	52	28	35	41	54	27	39

→

3. Arrange the following numbers in order from smallest to greatest.

101	10	110	1	11	100	111	0

→

4. Arrange the following numbers in order from smallest to greatest.

23	31	37	27	33	25	35	29

⟶

5. Arrange the following numbers in order from smallest to greatest.

75	125	80	115	90	120	85	100

⟶

6. Arrange the following numbers in order from smallest to greatest.

321	132	231	23	312	213	32	123

⟶

7. Arrange the following numbers in order from smallest to greatest.

107	153	142	149	125	170	169	114

8. Arrange the following numbers in order from smallest to greatest.

35	28	56	7	21	42	14	49

9. Arrange the following numbers in order from smallest to greatest.

301	288	310	300	299	219	309	298

10. Arrange the following numbers in order from smallest to greatest.

100	91	79	88	97	82	94	85

11. Arrange the following numbers in order from smallest to greatest.

42	351	579	24	975	468	135	864

12. Arrange the following numbers in order from smallest to greatest.

855	477	669	585	744	696	558	447

12

ALPHABETIZING
LISTS

first second third

Example: Alphabetize the following list.

monkey	*koala*	
lion	*leopard*	
koala	*lion*	
mouse	*monkey*	
leopard	*mouse*	

Solution:

- Look at the first letters. Koala begins with k, lion and leopard begin with l, and monkey and mouse begin with m. Koala comes first because k comes before l and m.

- Next look at the second letters. Leopard comes before lion because e comes before i.

- Next look at the third letters. Monkey comes before mouse because n comes before u.

- Enter the words in order in the right column. (We did this step for you in this example.)

1. Alphabetize the following list.

meow	
gobble	
moo	
buzz	
oink	
baa	

2. Alphabetize the following list.

sun	
rain	
snow	
cloud	
thunder	
lightning	

3. Alphabetize the following list.

daughter	
niece	
grandma	
son	
nephew	
grandpa	

4. Alphabetize the following list.

salt	
pepper	
milk	
cereal	
peas	
corn	

5. Alphabetize the following list.

stapler	
pencil	
eraser	
paper	
scissors	
glue	

6. Alphabetize the following list.

tennis	
swimming	
basketball	
hockey	
baseball	
soccer	

7. Alphabetize the following list.

Why	
Who	
How	
When	
What	
Where	

8. Alphabetize the following list.

tough	
though	
through	
thorough	
thought	
throughout	

13

WHICH ACTION COMES FIRST?

Before and After

Each exercise gives you two actions. Circle the action that comes first.

Example:

- put shoes on
- put socks on

You have to put socks on before you can put shoes on. Therefore, "put socks on" comes first. Circle this answer. (We did this step for you in this example.)

1. Circle the action that comes first.

   - eat a pizza

   - order a pizza

2. Circle the action that comes first.

   - pour milk into a bowl

   - open the refrigerator

3. Circle the action that comes first.

   - go to the rest room

   - flush the toilet

4. Circle the action that comes first.

   - a flower grows

   - a seed is plantcd

5. Circle the action that comes first.

- unlock a door

- open the door

6. Circle the action that comes first.

- seal an envelope

- mail the letter

7. Circle the action that comes first.

- play a baseball game

- decide on the teams

8. Circle the action that comes first.

- chew gum

- blow a bubble

9. Circle the action that comes first.

- fall asleep

- have sweet dreams

10. Circle the action that comes first.

- watch a play

- clap your hands

11. Circle the action that comes first.

- eat ice-cream

- use a napkin

12. Circle the action that comes first.

- read page 42

- open a book

13. Circle the action that comes first.

- become a teacher

- learn how to do something

14. Circle the action that comes first.

- dance to music

- turn on the radio

15. Circle the action that comes first.

- save your money

- break your piggy bank open

16. Circle the action that comes first.

- a butterfly forms

- a caterpillar forms

14

PATTERN PUZZLES

2, 5, 8, 11, 14, 17, 20, 23

1. Fill in the two blanks to complete the pattern.

B	D	F	H	J		N	

2. Fill in the two blanks to complete the pattern.

5	10	15		25	30		40

3. Fill in the two blanks to complete the pattern.

Y	X	W	V		T	S	

4. Fill in the two blanks to complete the pattern.

16	14	12		8		4	2

5. Fill in the two blanks to complete the pattern.

☺	☹	☺	☹	☺			☹

6. Fill in the two blanks to complete the pattern.

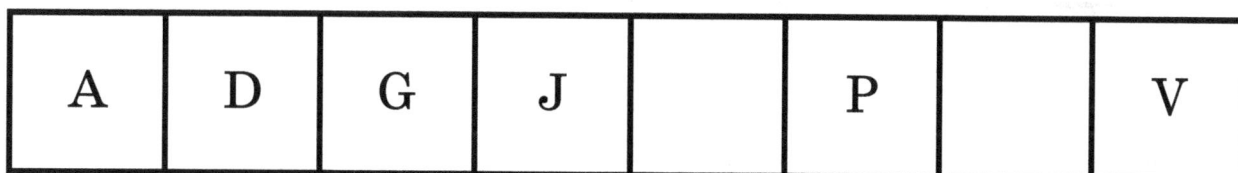

A	D	G	J		P		V

7. Fill in the two blanks to complete the pattern.

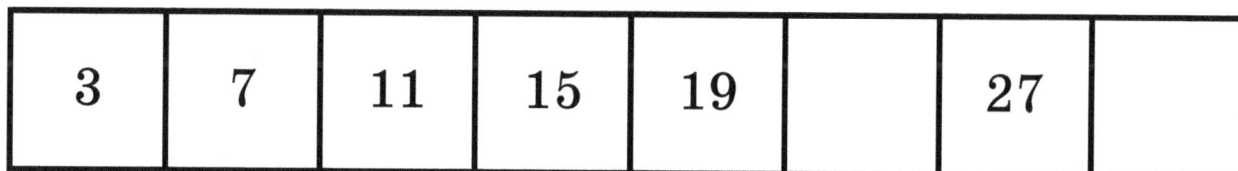

3	7	11	15	19		27	

8. Fill in the two blanks to complete the pattern.

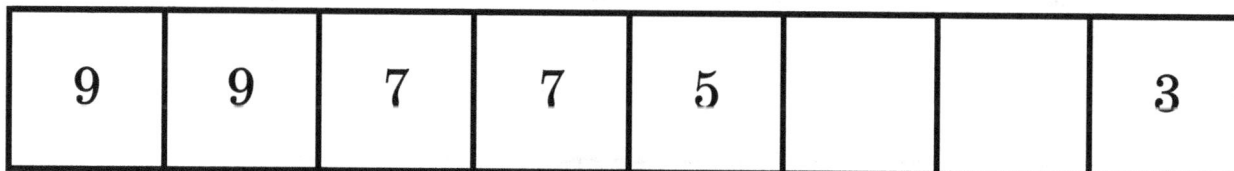

9	9	7	7	5			3

9. Fill in the two blanks to complete the pattern.

W	U	S		O		K	I

10. Fill in the two blanks to complete the pattern.

93	89	85	81		73		65

11. Fill in the two blanks to complete the pattern.

140	130	120		100		80	70

12. Fill in the two blanks to complete the pattern.

1	2	3	5	8	13		

13. Fill in the two blanks to complete the pattern.

1	2	4	8	16			128

14. Fill in the two blanks to complete the pattern.

1234		1324	1342		1432	2134	2143

15. Fill in the two blanks to complete the pattern.

3	4	6	7	9	10		

16. Fill in the two blanks to complete the pattern.

A9	C8	E7		I5		M3	O2

17. Fill in the two blanks to complete the pattern.

63	56	49		35		21	14

18. Fill in the two blanks to complete the pattern.

B	C	E	F	H	I		

19. Fill in the two blanks to complete the pattern.

→	↑	←	↓	→	↑		

20. Fill in the two blanks to complete the pattern.

15

THINKING
PUZZLES

Christmas Day = December 25

New Year's Day = January 1

Valentine's Day = February 14

Leap Year = February 29

Note: A date includes a month and a day, such as September 14 or August 21. See the previous page for a few notable dates.

1. What is the date of the day before Christmas?

2. What is the date of the day after Valentine's Day?

3. What is the date of the day after Leap Year?

4. What is the date of the day before New Year's?

Tip: It may help to review Chapter **6.**

5. If today is Tuesday, which day is tomorrow?

6. If today is Sunday, which day was yesterday?

7. If tomorrow is Thursday, which day is today?

8. If yesterday was Monday, which day is today?

9. If tomorrow is Friday, which day was yesterday?

10. Given that P is the 16th letter of the alphabet, what is the 18th letter?

11. Given that L is the 12th letter of the alphabet, what is the 14th letter?

12. Given that Y is the 25th letter of the alphabet, what is the 24th letter?

13. Given that J is the 10th letter of the alphabet, what is the 8th letter?

14. Given that W is the 23rd letter of the alphabet, what is the 21st letter?

15. Tracy has $28. Bill gives $3 to Tracy. How much money does Tracy have now?

16. Doug has $31. Dough buys a snack for $2. How much money does Doug have now?

17. Carla has $57. Carla receives $3 from her mom. How much money does Carla have now?

18. Jeff has $40. Jeff spends $4 at the store. How much money does Jeff have now?

19. Anna has $80. Anna loses $1. How much money does Anna have now?

penny	nickel	dime	quarter	dollar
1 cent	5 cents	10 cents	25 cents	100 cents

20. Sam has 20 cents. Sam gives 1 penny to Dan. How much money does Sam have now?

21. Julie has 45 cents. Julie finds a nickel. How much money does Julie have now?

22. Jose has 3 quarters. Jose receives another quarter from his dad. How much money does Jose have now?

23. Elle has 72 cents. Elle throws 3 pennies into a wishing well. How much money does Elle have now?

ANSWER
KEY

Chapter 1: Uppercase Letters A to Z

Page 6:
1. F (comes after E)
2. S (comes after R)
3. K (comes after J)
4. Y (comes after X)
5. D (comes after C)
6. O (comes after N)
7. J (comes after I)
8. V (comes after U)

Page 7:
9. Q (comes after P)
10. E (comes after D)
11. L (comes after K)
12. Z (comes after Y)
13. T (comes after S)
14. N (comes after M)
15. W (comes after V)
16. H (comes after G)

Page 8:
17. P (comes after O)
18. G (comes after F)
19. X (comes after W)
20. I (comes after H)
21. U (comes after T)
22. R (comes after Q)
23. C (comes after B)
24. M (comes after L)

Note that pages 9-10 ask which letter comes "before" instead of "after."

Page 9:
25. W (comes before X)
26. F (comes before G)
27. O (comes before P)
28. I (comes before J)
29. U (comes before V)
30. R (comes before S)
31. M (comes before N)
32. Y (comes before Z)

Page 10:
33. Q (comes before R)
34. B (comes before C)
35. L (comes before M)
36. G (comes before H)
37. K (comes before L)
38. S (comes before T)
39. H (comes before I)
40. X (comes before Y)

Chapter 2: Numbers 1–100

Page 12:

    1. 7 (comes after 6)

    2. 43 (comes after 42)

    3. 92 (comes after 91)

    4. 18 (comes after 17)

    5. 30 (comes after 29)

    6. 79 (comes after 78)

    7. 56 (comes after 55)

    8. 61 (comes after 60)

Page 13:

    9. 15 (comes after 14)

    10. 40 (comes after 39)

    11. 88 (comes after 87)

    12. 69 (comes after 68)

    13. 41 (comes after 40)

    14. 54 (comes after 53)

    15. 26 (comes after 25)

    16. 100 (comes after 99)

Page 14:

    17. 64 (comes after 63)

    18. 45 (comes after 44)

    19. 57 (comes after 56)

    20. 20 (comes after 19)

    21. 29 (comes after 28)

    22. 72 (comes after 71)

    23. 36 (comes after 35)

    24. 93 (comes after 92)

Note that pages 15-16 ask which number comes "before" instead of "after."

Page 15:

    25. 7 (comes before 8)

    26. 31 (comes before 32)

    27. 76 (comes before 77)

    28. 25 (comes before 26)

    29. 48 (comes before 49)

    30. 59 (comes before 60)

    31. 80 (comes before 81)

    32. 12 (comes before 13)

Page 16:

    33. 60 (comes before 61)

    34. 29 (comes before 30)

    35. 88 (comes before 89)

    36. 47 (comes before 48)

    37. 26 (comes before 27)

    38. 99 (comes before 100)

    39. 5 (comes before 6)

    40. 74 (comes before 75)

Chapter 3: Lowercase Letters a to z

Page 18:

1. l (comes after k)
2. e (comes after d)
3. z (comes after y)
4. k (comes after j)
5. p (comes after o)
6. g (comes after f)
7. b (comes after a)
8. r (comes after q)

Page 19:

9. m (comes after l)
10. h (comes after g)
11. q (comes after p)
12. y (comes after x)
13. c (comes after b)
14. t (comes after s)
15. j (comes after i)
16. u (comes after t)

Page 20:

17. d (comes after c)
18. s (comes after r)
19. i (comes after h)
20. o (comes after n)
21. v (comes after u)
22. x (comes after w)
23. n (comes after m)
24. f (comes after e)

Note that pages 21-22 ask which letter comes "before" instead of "after."

Page 21:

25. u (comes before v)
26. j (comes before k)
27. r (comes before s)
28. h (comes before i)
29. f (comes before g)
30. x (comes before y)
31. c (comes before d)
32. n (comes before o)

Page 22:

33. e (comes before f)
34. s (comes before t)
35. o (comes before p)
36. k (comes before l)
37. p (comes before q)
38. i (comes before j)
39. y (comes before z)
40. w (comes before x)

Chapter 4: Big Number Challenge (100+)

Page 24:

1. 124 (comes after 123)
2. 655 (comes after 654)
3. 1,208 (comes after 1,207)
4. 257 (comes after 256)
5. 200 (comes after 199)
6. 349 (comes after 348)
7. 301 (comes after 300)
8. 110 (comes after 109)

Page 25:

9. 946 (comes after 945)
10. 130 (comes after 129)
11. 476 (comes after 475)
12. 701 (comes after 700)
13. 334 (comes after 333)
14. 812 (comes after 811)
15. 600 (comes after 599)
16. 2,058 (comes after 2,057)

Page 26:

17. 145 (comes after 144)
18. 720 (comes after 719)
19. 3,001 (comes after 3,000)
20. 329 (comes after 328)
21. 531 (comes after 530)
22. 10,000 (comes after 9,999)
23. 2,000 (comes after 1,999)
24. 1,100 (comes after 1,099)

Note that pages 27-28 ask which number comes "before" instead of "after."

Page 27:

25. 242 (comes before 243)
26. 898 (comes before 899)
27. 527 (comes before 528)
28. 299 (comes before 300)
29. 1,233 (comes before 1,234)
30. 776 (comes before 777)
31. 3,999 (comes before 4,000)
32. 325 (comes before 326)

Page 28:

33. 109 (comes before 110)
34. 477 (comes before 478)
35. 575 (comes before 576)
36. 269 (comes before 270)
37. 218 (comes before 219)
38. 99 (comes before 100)
39. 986 (comes before 987)
40. 1,499 (comes before 1,500)

Chapter 5: Number Words

Page 30:

1. eight (8) comes after seven (7)
2. fifty-seven (57) comes after fifty-six (56)
3. twenty-one (21) comes after twenty (20)
4. ninety-six (96) comes after ninety-five (95)
5. nineteen (19) comes after eighteen (18)
6. twelve (12) comes after eleven (11)
7. forty-four (44) comes after forty-three (43)
8. forty (40) comes after thirty-nine (39)

Page 31:

9. seventeen (17) comes after sixteen (16)
10. thirty-six (36) comes after thirty-five (35)
11. ninety-three (93) comes after ninety-two (92)
12. seventy-two (72) comes after seventy-one (71)
13. forty-eight (48) comes after forty-seven (47)
14. twenty (20) comes after nineteen (19)
15. fifty-nine (59) comes after fifty-eight (58)
16. twenty-five (25) comes after twenty-four (24)

Page 32:

17. thirty-five (35) comes after thirty-four (34)
18. seven (7) comes after six (6)
19. twenty-three (23) comes after twenty-two (22)
20. seventy-seven (77) comes after seventy-six (76)
21. fifty-one (51) comes after fifty (50)
22. ninety-nine (99) comes after ninety-eight (98)
23. fourteen (14) comes after thirteen (13)
24. eighty (80) comes after seventy-nine (79)

Note that pages 33-34 ask which number comes "before" instead of "after."

Page 33:

25. eight (8) comes before nine (9)

26. eighty-nine (89) comes before ninety (90)

27. sixty-two (62) comes before sixty-three (63)

28. ten (10) comes before eleven (11)

29. twenty-seven (27) comes before twenty-eight (28)

30. nine (9) comes before ten (10)

31. eighty-six (86) comes before eighty-seven (87)

32. thirty-five (35) comes before thirty-six (36)

Page 34:

33. forty-one (41) comes before forty-two (42)

34. eleven (11) comes before twelve (12)

35. seventy-two (72) comes before seventy-three (73)

36. twenty-nine (29) comes before thirty (30)

37. six (6) comes before seven (7)

38. fifty (50) comes before fifty-one (51)

39. ninety-nine (99) comes before one hundred (100)

40. zero (0) is the best answer for which number comes before one (1)

Chapter 6: Days of the Week

Page 37:

    1. Wednesday (comes after Tuesday)

    2. Saturday (comes after Friday)

    3. Friday (comes after Thursday)

    4. Monday (comes after Sunday)

    5. Thursday (comes after Wednesday)

    6. Tuesday (comes after Monday)

    7. Sunday (comes after Saturday)

Page 38:

    8. Saturday (comes after Friday)

    9. Monday (comes after Sunday)

    10. Friday (comes after Thursday)

    11. Sunday (comes after Saturday)

    12. Thursday (comes after Wednesday)

    13. Wednesday (comes after Tuesday)

    14. Tuesday (comes after Monday)

Note that pages 39-40 ask which day comes "before" instead of "after."

Page 39:

    15. Tuesday (comes before Wednesday)

    16. Sunday (comes before Monday)

    17. Wednesday (comes before Thursday)

    18. Saturday (comes before Sunday)

    19. Thursday (comes before Friday)

    20. Monday (comes before Tuesday)

    21. Friday (comes before Saturday)

Page 40:

    22. Wednesday (comes before Thursday)

    23. Sunday (comes before Monday)

    24. Friday (comes before Saturday)

    25. Monday (comes before Tuesday)

    26. Thursday (comes before Friday)

    27. Saturday (comes before Sunday)

    28. Tuesday (comes before Wednesday)

Chapter 7: Months of the Year

Page 43:

1. April (comes after March)
2. July (comes after June)
3. October (comes after September)
4. March (comes after February)
5. August (comes after July)
6. September (comes after August)
7. December (comes after November)
8. June (comes after May)

Page 44:

9. February (comes after January)
10. November (comes after October)
11. May (comes after April)
12. January (comes after December)
13. March (comes after February)
14. August (comes after July)
15. October (comes after September)
16. September (comes after August)

Note that pages 45-46 ask which month comes "before" instead of "after."

Page 45:

    17. May (comes before June)

    18. October (comes before November)

    19. January (comes before February)

    20. April (comes before May)

    21. December (comes before January)

    22. March (comes before April)

    23. November (comes before December)

    24. February (comes before March)

Page 46:

    25. June (comes before July)

    26. March (comes before April)

    27. August (comes before September)

    28. October (comes before November)

    29. July (comes before August)

    30. February (comes before March)

    31. September (comes before October)

    32. April (comes before May)

Chapter 8: Colors of the Rainbow

Page 49:

    1. yellow (comes after orange)

    2. blue (comes after green)

    3. green (comes after yellow)

    4. red (comes at the top of the rainbow)

    5. violet (comes after indigo)

    6. orange (comes after red)

    7. indigo (comes after blue)

Page 50:

    8. green (comes after yellow)

    9. yellow (comes after orange)

    10. blue (comes after green)

    11. orange (comes after red)

    12. violet (comes after indigo)

    13. red (comes at the top of the rainbow)

    14. indigo (comes after blue)

Note that pages 51-52 ask which color comes "before" instead of "after."

Also note that questions 21 and 24 ask which color comes at the "bottom" instead of the "top."

Page 51:

    15. yellow (comes before green)

    16. indigo (comes before violet)

    17. green (comes before blue)

    18. orange (comes before yellow)

    19. blue (comes before indigo)

    20. red (comes before orange)

    21. violet (comes at the bottom of the rainbow)

Page 52:

    22. blue (comes before indigo)

    23. green (comes before blue)

    24. violet (comes at the bottom of the rainbow)

    25. orange (comes before yellow)

    26. indigo (comes before violet)

    27. red (comes before orange)

    28. yellow (comes before green)

Chapter 9: Planets of the Solar System

Page 55:

1. Mars (comes after Earth)
2. Uranus (comes after Saturn)
3. Venus (comes after Mercury)
4. Saturn (comes after Jupiter)
5. Neptune (comes after Uranus)
6. Earth (comes after Venus)
7. Jupiter (comes after Mars)
8. Mercury (is closest to the sun)

Page 56:

9. Jupiter (comes after Mars)
10. Saturn (comes after Jupiter)
11. Earth (comes after Venus)
12. Mercury (is closest to the sun)
13. Mars (comes after Earth)
14. Neptune (comes after Uranus)
15. Venus (comes after Mercury)
16. Uranus (comes after Saturn)

Note that pages 57-58 ask which planet comes "before" instead of "after."

Also note that questions 19 and 30 ask which planet is "furthest" instead of "closest" to the sun.

Page 57:
    17. Mars (comes before Jupiter)
    18. Mercury (comes before Venus)
    19. Neptune (is furthest from the sun)
    20. Saturn (comes before Uranus)
    21. Earth (comes before Mars)
    22. Jupiter (comes before Saturn)
    23. Venus (comes before Earth)
    24. Uranus (comes before Neptune)

Page 58:
    25. Jupiter (comes before Saturn)
    26. Uranus (comes before Neptune)
    27. Venus (comes before Earth)
    28. Mars (comes before Jupiter)
    29. Mercury (comes before Venus)
    30. Neptune (is furthest from the sun)
    31. Earth (comes before Mars)
    32. Saturn (comes before Uranus)

Chapter 10: Polygons (Shapes with Straight Edges)

Page 61:

1. pentagon ⬠ 5 (has one more side than a square ☐ 4)

2. heptagon   7   (has one more side than a hexagon   6)

3. square ☐ 4 (has one more side than a triangle △ 3)

4. hexagon   6   (has one more side than a pentagon   5)

5. octagon   8   (has one more side than a heptagon   7)

Page 62:

6. decagon   10   (has one more side than a nonagon   9)

7. square ☐ 4 (has one more side than a triangle △ 3)

8. nonagon   9   (has one more side than an octagon   8)

9. hexagon   6   (has one more side than a pentagon   5)

10. pentagon   5   (has one more side than a square ☐ 4)

Note that pages 63-64 ask which polygon has one "less" side instead of one "more."

Page 63:

11. pentagon 5 (has one less side than a hexagon 6)

12. triangle 3 (has one less side than a square 4)

13. octagon 8 (has one less side than a nonagon 9)

14. square 4 (has one less side than a pentagon 5)

15. hexagon 6 (has one less side than a heptagon 7)

Page 64:

16. heptagon 7 (has one less side than an octagon 8)

17. nonagon 9 (has one less side than a decagon 10)

18. square 4 (has one less side than a pentagon 5)

19. triangle 3 (has one less side than a square 4)

20. pentagon 5 (has one less side than a hexagon 6)

Chapter 11: Sorting Numbers

Page 67:

1.

8	13	7	2	9	16	4	15
2	4	7	8	9	13	15	16

2.

43	52	28	35	41	54	27	39
27	28	35	39	41	43	52	54

3.

101	10	110	1	11	100	111	0
0	1	10	11	100	101	110	111

Page 68:

4.

23	31	37	27	33	25	35	29
23	25	27	29	31	33	35	37

5.

75	125	80	115	90	120	85	100
75	80	85	90	100	115	120	125

6.

321	132	231	23	312	213	32	123
23	32	123	132	213	231	312	321

Page 69:

7.

107	153	142	149	125	170	169	114
107	114	125	142	149	153	169	170

8.

35	28	56	7	21	42	14	49
7	14	21	28	35	42	49	56

9.

301	288	310	300	299	219	309	298
219	288	298	299	300	301	309	310

Page 70:

10.

100	91	79	88	97	82	94	85
79	82	85	88	91	94	97	100

→

11.

42	351	579	24	975	468	135	864
24	42	135	351	468	579	864	975

→

12.

855	477	669	585	744	696	558	447
447	477	558	585	669	696	744	855

→

Chapter 12: Alphabetizing Lists

Page 73:

1.

meow	baa
gobble	buzz
moo	gobble
buzz	meow
oink	moo
baa	oink

2.

sun	cloud
rain	lightning
snow	rain
cloud	snow
thunder	sun
lightning	thunder

Page 74:

3.

daughter	daughter
niece	grandma
grandma	grandpa
son	nephew
nephew	niece
grandpa	son

4.

salt	cereal
pepper	corn
milk	milk
cereal	peas
peas	pepper
corn	salt

Page 75:

5.

stapler	eraser
pencil	glue
eraser	paper
paper	pencil
scissors	scissors
glue	stapler

6.

tennis	baseball
swimming	basketball
basketball	hockey
hockey	soccer
baseball	swimming
soccer	tennis

Page 76:

7.

Why	How
Who	What
How	When
When	Where
What	Who
Where	Why

8.

tough	thorough
though	though
through	thought
thorough	through
thought	throughout
throughout	tough

Chapter 13: Which Action Comes First?

Page 79:

1. You order a pizza before you eat it.

- eat a pizza
- order a pizza

2. You open the refrigerator to get the milk before you pour it into a bowl.

- pour milk into a bowl
- open the refrigerator

3. You go to the rest room before you flush the toilet.

- go to the rest room
- flush the toilet

4. You plant a seed before the flower grows.

- a flower grows
- a seed is planted

Page 80:

5. Unlock the door before you open it.

- unlock a door
- open the door

6. You seal an envelope before you mail the letter.

- seal an envelope
- mail the letter

7. You decide on teams before you play a baseball game.

- play a baseball game
- decide on the teams

8. Once you start chewing gum, you can blow a bubble with it.

- chew gum
- blow a bubble

Page 81:

9. You fall asleep before you dream.

- fall asleep
- have sweet dreams

10. If you watch a play, you clap your hands when it is finished.

- watch a play
- clap your hands

11. After you eat ice-cream, you may need to wipe your face and hands with a napkin.

- eat ice-cream
- use a napkin

12. You must open a book in order to read any of its pages.

- read page 42
- open a book

Page 82:

13. You should learn how to do something well before you attempt to teach it.

- become a teacher
- learn how to do something

14. First turn on the radio so that there will be music to dance along to.

- dance to music
- turn on the radio

15. Save money in a piggy bank before you break it open.

- save your money
- break your piggy bank open

16. A caterpillar will turn into a butterfly. (This is called metamorphosis.)

- a butterfly forms
- a caterpillar forms

Chapter 14: Pattern Puzzles

Page 84:

1. Skip every other letter: B (skip C) D (skip E) F (skip G) H (skip I) J (skip K) L (skip M) N (skip O) P.

B	D	F	H	J	L	N	P

2. Count by 5.

5	10	15	20	25	30	35	40

3. Write the alphabet right to left starting with R.

Y	X	W	V	U	T	S	R

4. Count down by 2.

16	14	12	10	8	6	4	2

Page 85:

5. Happy, sad, happy, sad, happy, sad, happy, sad.

☺	☹	☺	☹	☺	☹	☺	☹

6. A (skip BC) D (skip EF) G (skip HI) J (skip KL) M (skip NO) P (skip QR) S (skip TU) V.

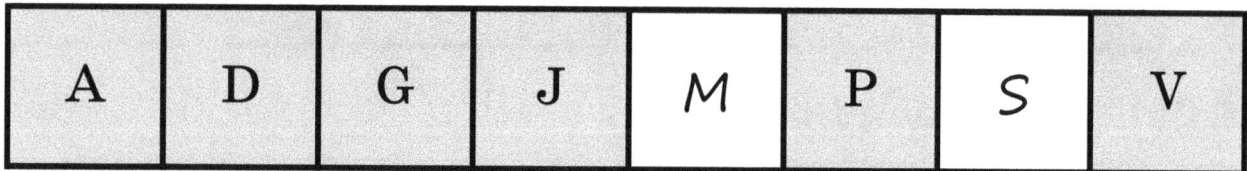

A	D	G	J	M	P	S	V

7. Count by 4.

3	7	11	15	19	23	27	31

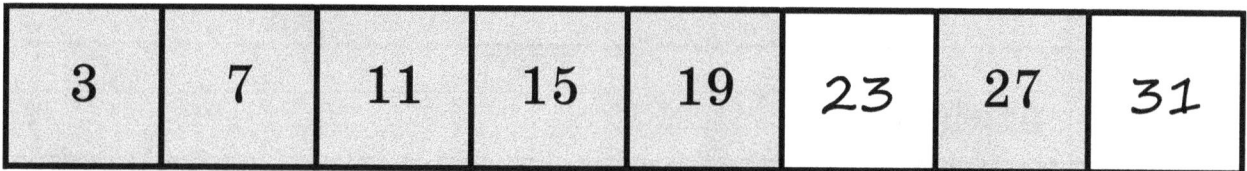

8. The pattern is 9, 7, 5, 3, except that each number is repeated to make 9, 9, 7, 7, 5, 5, 3, 3.

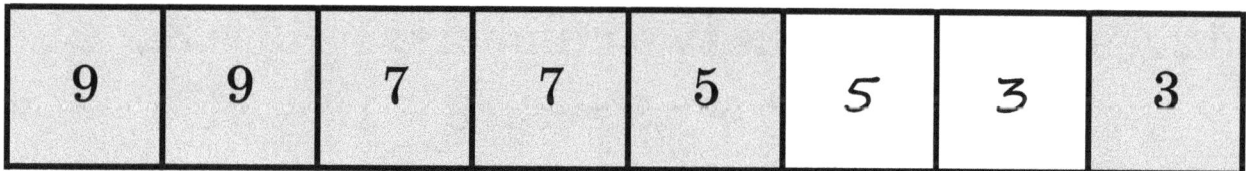

9	9	7	7	5	5	3	3

Page 86:

9. W (skip V) U (skip T) S (skip R) Q (skip P) O (skip N) M (skip L) K (skip J) I.
(Note: You don't have to think backwards. You can go forwards from right to left.)

W	U	S	Q	O	M	K	I

10. Count down by 4.

93	89	85	81	77	73	69	65

11. Count down by 10.

140	130	120	110	100	90	80	70

12. Add the previous two numbers: $1 + 2 = 3$, $2 + 3 = 5$, $3 + 5 = 8$, $5 + 8 = 13$, $8 + 13 = 21$, $13 + 21 = 34$.

1	2	3	5	8	13	21	34

Page 87:

13. Double the previous number: 2 is twice 1, 4 is twice 2, 8 is twice 4, 16 is twice 8, and so on.

1	2	4	8	16	32	64	128

14. These are the different ways to order the digits 1, 2, 3, and 4 in numerical order.

1234	1243	1324	1342	1423	1432	2134	2143

15. 3, 4 (skip 5), 6, 7 (skip 8), 9, 10 (skip 11), 12, 13.

3	4	6	7	9	10	12	13

16. A (skip B) C (skip D) E (skip F) G (skip H) I (skip J) K (skip L) M (skip N) O.
Also, count down by 1.

A9	C8	E7	G6	I5	K4	M3	O2

Page 88:

17. Count down by 7 (or count up by 7 from right to left).

63	56	49	42	35	28	21	14

18. B, C (skip D) E, F (skip G) H, I (skip J) K, L.

B	C	E	F	H	I	K	L

19. Rotate each arrow one quarter turn counterclockwise (or repeat the pattern right, up, left, down).

→	↑	←	↓	→	↑	←	↓

20. Draw a polygon with one less side (we learned about polygons in Chapter 10).

10	9	8	7	6	5	4	3

Chapter 15: Thinking Puzzles

Page 90:

    1. December 24 (comes <u>before</u> December 25)

      (Note that December 24 is Christmas Eve.)

    2. February 15 (comes <u>after</u> February 14)

    3. March 1 (comes <u>after</u> February 29)

      (Note that Leap Year only comes once every four years.)

    4. December 31 (comes <u>before</u> January 1)

      (Note that December 31 is New Year's Eve.)

Page 91:

    5. Wednesday (is the day after Tuesday)

    6. Saturday (is the day before Sunday)

    7. Wednesday (is the day before Thursday)

    8. Tuesday (is the day after Monday)

    9. Wednesday (since today is Thursday, yesterday is Wednesday)

Page 92:

    10. R (comes 2 letters after P)

    11. N (comes 2 letters after L)

    12. X (is the letter before Y)

    13. H (comes 2 letters before J)

    14. U (comes 2 letters before W)

Page 93:

    15. $31 (is $3 more than $28)

    16. $29 (is $2 less than $31)

    17. $60 (is $3 more than $57)

    18. $36 (is $4 less than $40)

    19. $79 (is $1 less than $80)

Page 94:

20. 19 cents (is 1 cent less than 20 cents)

21. 50 cents (is 5 cents more than 45 cents)

22. 1 dollar (is equal to 4 quarters)

   (Note that 100 cents is the same as 1 dollar.)

23. 69 cents (is 3 cents less than 72 cents)

CURSIVE HANDWRITING

It's never too late to learn cursive handwriting.

- Learn how to write the cursive alphabet.

- Practice writing words, phrases, and sentences.

- Challenge yourself to remember how to write each letter in cursive.

- Writing prompts offer additional practice.

The Art of *Cursive* Handwriting

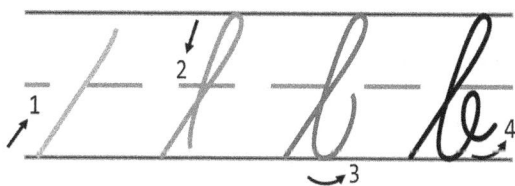

Jenny Pearson

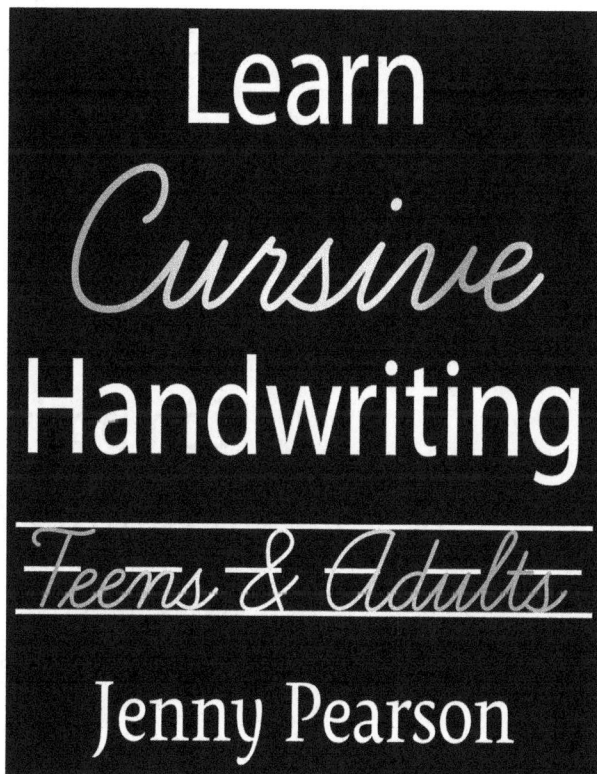

Learn *Cursive* Handwriting

Teens & Adults

Jenny Pearson

EXPRESS YOURSELF

Practice your writing skills with this composition writing prompts workbook. This book includes:

- a healthy variety of writing prompts.

- literary devices like similes and metaphors.

- a good mix of short answers with composition practice so as not to seem too intimidating.

- different forms of writing, such as opinions, descriptions, interviews, and short stories.

EXPRESS YOURSELF
Composition Writing Prompts Workbook

ideas • emotions • memories • issues
feelings • opinions • goals • thoughts
hopes • dreams • senses • vision

Jenny Pearson

SPELLING AND PHONICS

Spelling and phonics go hand-in-hand together:

- In *The Art of Spelling*, you learn techniques for how to spell a word after you've heard it spoken.

- In *The Art of Phonics*, you learn techniques for how to pronounce a word that you see in writing.

The Art of **PHONICS**	
tough	through
though	thorough
thought	throughout

Jenny Pearson

The Art of **SPELLING**		
s	spag	spaghet
sp	spagh	spaghett
spa	spaghe	spaghetti

Jenny Pearson

COLORING BOOKS

Coloring books aren't just for kids. They are popular among teens and adults, too. Coloring provides a relaxing way to take your mind off of stress, and lets you use your creativity.

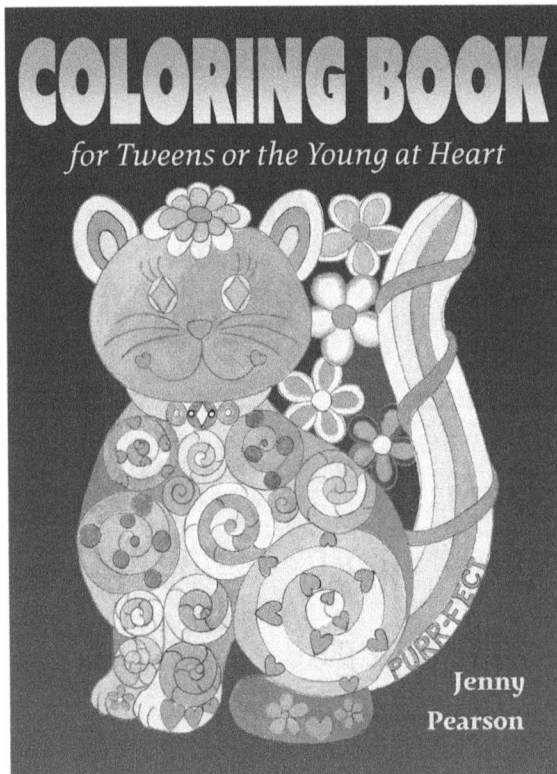

COLORING BOOK
for Tweens or the Young at Heart

Jenny Pearson

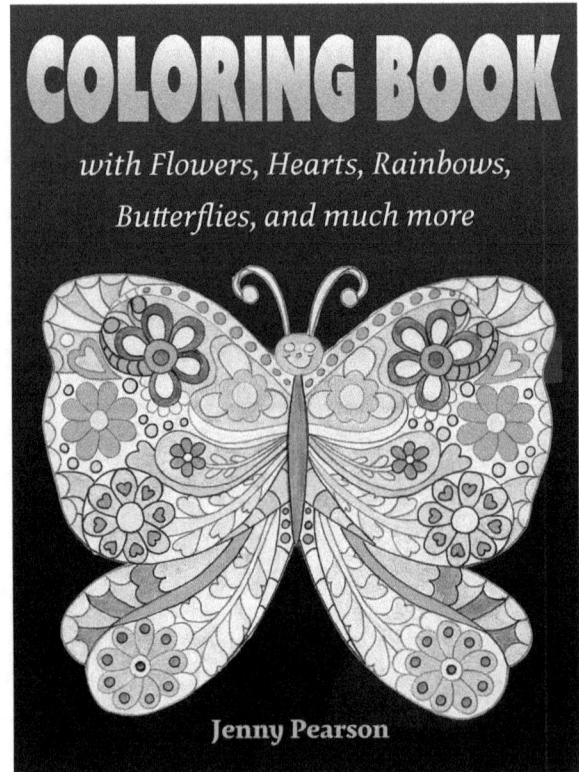

COLORING BOOK
with Flowers, Hearts, Rainbows, Butterflies, and much more

Jenny Pearson